# Cuban Quest for Brain Implant Truth

By Chad Kister
Copyright 2019 by Chad Kister

## Table of Contents

## Chapter 1: Knocked out and Brain Implanted in my House

I heard a knock on my door. I answered, and one of my tenants, Ross Martin knocked me out with a baseball bat.

I awoke on March 28, 2009 in my bed, with severe pain in my right temple, and hearing the thoughts of my two tenants at the same time, Sasha Sigetic and Ross Martin. They stole my thoughts. My life would never be the same. (See the movies, The Speed of Thought and American Ultra).

I was the author of 4 books, and have given more than a thousand speeches around the world. Now, I would be considered a lunatic for the rest of my life, unless they admit to the truth about what they did to me, and their use of brain implant technology in my head.

I had overheard them talking about having a contract with the Department of Homeland Security to search my house. But they were frustrated that one of my wireless cameras broadcast to where others could see and hear it. I was horrified as I thought that was a private camera. I set up a meeting with 2 FBI agents, and secretly recorded it, complaining about their breaking and entering. Then, they hit me with a baseball bat, and put in a brain implant, and life, would never be the same...

I noticed that something seemed wrong with my dog, New Moon. She seemed to have a head injury, and was acting very odd.

At first I thought they were playing sounds, through the register. I went outside. They mind controlled me, thinking "go to Krogers." But instead of just thinking it in their head, the mind controllers (gang stalking government terrorists) thought it into my head, hoping and initially being successful in mind controlling me.

I drove over to Krogers. Then I was told to go back. I turned my car around in a driveway, and I went back. Then they said go back to Krogers. I turned around in my driveway, and an SUV nearly hit me.

I then went to Krogers. At Krogers, I bought coffee, then returned.

My tenants, Sasha Sigetic and Ross Martin, were now connected by thought to my head. Because so few realize that this technology exists, so many question my sanity. Many others are simply controlled by what the thought police demand, and do not realize it or are too afraid to speak out about it or they will be locked up, robbed, and lose everything.

I still thought they must just be playing sounds near me. I walked down to the building I call Bessemer Place. I thought there I could escape the sound. But there was no escape, it went into a cochlear implant implanted in my head, wired to a braingate 2 wireless implant in my forehead, attached to a satellite transceiving device. Is this the "roving wiretap" under the Patriot Act?

Then I went on a walk, a long walk back in the forest. I could not believe I could still hear their thoughts, and that there was no escape for the rest of my life. I continued to walk back, miles into the forest, on the ORV trails. This was why I moved to Bessemer Rd., near Nelsonville: to be able to hike miles into the forest. After many miles, I turned around and went back.

## Chapter 2: A Mission to Stop Climate Change

I had a mission: I was editing my 4th book, Arctic Screaming: Journey to the Front Line of the Climate Crisis. It warned of massive harm by climate change through flooding, droughts, diseases, tornadoes, hurricanes and other severe weather.

Back at my house, I continued my efforts to finish my 4th book. I found editing very difficult with the thought distractions. They "thought injected" that they had poisoned my food. I wondered rather to throw my precious food away or eat it. I learned early that Sasha and Ross were liars and not to believed. They were hired by someone to cause psychological terrorism, and mind control and they have done a very good job. It appears the fossil fuel industry, medical and government thugs, in one giant conspiracy have managed to mind control their critics. Until they admit to it, I will not know exactly who (except those named) and why I can just report what I know.

I had a major crisis: someone had implanted me at the same time I had a deadline to finish my book. I worked hard on editing the book. Sasha and Ross mind controlled me to drive to the Logan hospital and ask them to give me measles shots.

Sasha thought injected "measles shots, measles shots, Chad needs his measles shots." This was my first day of brain implant Hell, and I was under their control.

I drove to the Logan hospital and went to the Emergency Room. I said that I needed Measles shots. After sitting a few hours, they came out and said that they were only given to children. I drove home. I had no idea what to do.

Between editing my book, I looked up how they could do this to me. I realized that they must have something in my brain. I looked it up on the internet, and found the words brain implant, neural implant, and brain-computer interface. I downloaded articles.

## Chapter 3: The Smoking Gun: X-Rays at OBleness Memorial Hospital

I decided to get an x-ray. I went to OBleness Memorial Hospital on April 13, 2009. They refused. I said I would not leave until they gave me an x-ray of my head. I sat for hours reading in the waiting room. Eventually, they agreed to do x-rays. They took a few from the side of my head. I had a large terbyte drive in a backpack, which they stored away from the x-rays. I knew they were trying to delete my files. They had already deleted three videos of tree sitters in the Zaleski state forest. I pointed out the implant in my forehead to the doctor.

The next day, I drove in and got my x-rays. I went to medical records, filled out the forms, and waited. They gave them to me on a CD.

Back at home, I put the CD into a PC Windows Vista laptop computer, and a virus crashed it. I used a linux computer running windows to extract the x-ray images, and immediately saw the braingate2 wireless implant in my forehead, and the Choclear in my right ear. I took screenshots to document it. I uploaded them onto the internet at my www.chadkister.com website, and contacted the local media by phone.

Next, I worked on a press release, announcing the documentation of the implant on the internet. I decided to announce it in third person, that I knew the person, and had the x-rays up on the net. I blacked out my Birthday on the x-rays to stay anonymous. I was afraid that I would look like a lunatic, if I said that it was me.

I thought it would be in the national media instantly. Instead, I just waited, while the media snoozed.

## Chapter 4: Brain Implant Causes Mahem

My publisher told me that he missed the deadline, and could not publish my book. I was shocked. He did allow me to take it elsewhere. I worked on sending query letters to other publishers.

My tenants were a menace. They were clearly here just to mess with me. They made me permanently disabled by government brain implant. I asked Ross Martin about the brain implant. "We really don't need it," he said. "With credit cards, and cell phones." He is trained in psychological terrorism, constantly thinking into my head that my Mom was dead, my brother got in a car accident, and whenever I leave, that my house was ransacked, or burned down.

They knew my fears and used them to torture. What a sadistic terrorist beast I was talking to. He was a pawn for his father, Kevin Martin, whom I had forced to pay me the minimum wage that I had labored and toiled so hard for. I had worked for him with Athens Lawn and Garden.

I filed a complaint with the Ohio Attorney General's after Kevin refused to pay nearly $1,000 of desperately needed hard-earned money working minimum wage landscaping in the hot sun. Kevin had vowed revenge, and forced me to sign a waiver that I would not sue him. The brain implants may be his revenge, connecting me to Ross and Sasha who were trying to mind control and use psychological terrorist techniques.

## Chapter 5: Press Release Gets Moderate Media Attention

I was shocked that the media did not jump on the story. I used Sendblaster to send tens of thousands of emails to the media. For the press release, I announced that two individuals had reported being implanted, saving for later that I was one of them:

**For Immediate Release: July 29, 2009 "Author: The Thought Police are Real in 2009"**

Two individuals in Ohio have reported having brain implants placed in them against their will. They report that they are being interrogated through their thoughts. While this sounds like science fiction, both of these individuals are extraordinarily credible. They are afraid because the motivation behind this is to lock them up as being insane, because they are civil liberties activists.

Before you think this is fiction, please look up "brain implant" on Google, and see for yourself. Also look up brain-computer interface. I have perused the latest science on this issue to confirm that this technology not only exists, but that the Department of Homeland Security and DARPA are spending billions of dollars on it.

This technology has been here for years, including uses to help paraplegics connect to prosthetic limbs, and even surf the internet. But it is now being used as a form of thought policing and mind control, according to two individuals whom I have interviewed extensively about this who have brain implants currently.

For four months, I have investigated this. The individual is extraordinarily fit mentally, and very successful at protesting civil liberties violations committed by law enforcement.

Armed with dozens of credible science, government and news organization reports, I am ready to go public with this story, as I continue to investigate it further. I invite other news media to join in on this investigation. At the very least, readers should know that this technology is indeed here. It is amazing that so few know about it: this is a real scoop.

I have done four months of intensive research on this, and have copies of the x-rays from one of the individuals. Experts in the field, including Jeff Stibel, Chairman of Braingate, confirm that this technology does exist.

I have posted a link to the xray (one regular sized and one enlarged) at:

www.safeclimateact.org/brain1.jpg

www.safeclimateact.org/brain2.jpg

One small newspaper in Texas did publish my article: "The Thought Police are Real:"

"I am working on documenting a story about two individuals who have been implanted with brain implants against their will that monitors their thoughts and injects the voices of two people into their heads. Believe me, I wondered whether this technology existed when I first found out about it. But I have done four months of intensive research on this, and have copies of the x-rays from one of the individuals. Experts in the field confirm that this technology does exist, and both of these individuals are extraordinarily credible, but are afraid because the motivation behind this is to lock them up as being insane, because they are civil liberties activists.

Calling all x-ray technicians: please look at the frontal lobe (just under the forehead) at a black object, about 3 mm by 7 mm, and call me at 740-753-3000 or email chadkister@gmail.com to confirm this. Here are two copies of the x-rays, one showing the full brain, one zoomed in to the frontal lobe, where there is a brain implant.

www.safeclimateact.org/brain1.jpg

www.safeclimateact.org/brain2.jpg


**Chapter 6: The Thought Police are Real**

I published the article, "Thought Police are Real in 2009:"

Imagine a world where one's thoughts are monitored, and police could broadcast voices and other people's thoughts into ones brain. What if such devices were to be surreptitiously placed into people's brain's against their will? What power would that grant those who had such information, which could be broadcast wirelessly through satellites? Using their thoughts they could try to control people as well.

Every action, thought and everything one looked at could suddenly become recorded and used to extort a lifelong of slavery and obedience to the power doing the monitoring, allowing for a Hitler-like dominance of a country, leading to World War III. The latest in nanotechnology science shows that such technology has arrived. More troubling, such devices have been used in two Ohio citizens against their will, according to two sources. There are likely many, many more cases.

"I will never forget it," said Al Smith, owner of Spy Depot in suburban Columbus , Ohio about a time when a woman came in, asking for him to detect wireless frequencies in her head. He said that his device went wild, and that he detected frequencies with a second wireless device, a spectrum analyzer.

She brought in x-rays of her head, and he downloaded them onto his computer.

"I can read those things," he said, referring to x-rays. "I spotted it right away."

He said later that it was about an inch long and a quarter inch in diameter. She had recently been in a hospital. He said about two weeks later, two men came in, asking for him to give an affidavit about the woman. He would not give any information, because she was a client.

Some time later, she came in with the same two men, and asked for him to write up an affidavit. He wrote and signed an affidavit confirming that he had detected a wireless frequency coming from her head. "They say in ten years everyone will have one," he warned.

In a separate case, an anonymous source, whom the author knows personally is very credible, reported that he had been implanted with a microchip that monitored and influenced his thoughts, and had people speaking into his brain. Technically it is their thoughts, not speech.

He reported that two people, a male and a female were talking in his head, hypnotizing him and trying to control him. He said that they were responding to his thoughts, what he was doing and what he was seeing as well. He said that they were trying to make him go crazy, saying one thing, then saying the opposite.

The source wanted to remain anonymous, because he said their goal was to try to declare him mentally unfit to stand trial when he uncovered that they had put a nanotechnology transmitter into his brain. He said that they were hypnotizing him to call law enforcement, to turn himself in for things that they falsely accused him of, and woke him up through the night.

Nanotechnology has been trucking ahead with little oversight, with well documented studies confirming that the technology is there to see what people see, as well as monitor the brain. In 2004, the Food and Drug Administration approved the implantation of microchips into the brains of Alzheimer's patients to help them deal with the loss of memory.

What if such devices could be made extraordinarily small, utilizing nanotechnology and carbon fibers, such as to be near undetectable? What if they used very high frequencies, above 600 megahertz, such as to be undetectable by most private investigators?

Such technology was called uberveillance by Michael G. Michael, from the University of Wollongong 's School of Information Systems technology (ninemsn, March 20, 2009). Such technology could record what a person saw, the person's movements, and even their thoughts, Dr. Michael said.

One nightclub in Spain began implanting its patrons with Verichip microchips, about the size of a grain of rice, so that they could purchase drinks without the need to bring cash or a credit card, CNN reported June 9, 2004.  The nightclub could just scan the chip, which can be implanted for about $150, and be used as a debit or credit card.

VeriChip began implanting people in Mexico with a tracking microchip in 2003, which can be used to contain medical information, and to track a person within 5 miles, according to the Associated Press, July 17, 2003. The company said that they were working on a system to use satellite technology that could track people who were kidnapped.  Of course this could also be used for more nefarious purposes.

If this is what private companies are manufacturing, what is our government doing with the billions of dollars already spent on nanotechnology? We need greater transparency to see just what the Defense Department and the Department of Homeland Security have developed with billions of dollars spent on nanotechnology and brain implants.  We must ensure that any devices created can be detectable, and is never implanted into someone without one's consent.

With two credible reports in Ohio showing that people have been microchipped against their will, Congress needs to begin investigating this issue.  What would it be like to have someone be able to see as one goes to the bathroom, changes clothes or had sexual intercourse.  What if they continued to speak during such times, reminding you that they are watching?

What would happen to a person's dignity and self-worth if everything one thinks about, or was hypnotized to think or say, was monitored and being used against them? What if memories of movies, video games and television programs could be used against someone as "proof" they did something that they did not? Reading a book might make them think that what was written in the book was what the person did. What would it do to someone to have an endless probe, 24-7 with no way to stop it?

We cannot rely on the FBI. Two FBI workers were caught spying on teenage girls as they tried on prom dresses for 90 minutes in Morgantown, West Virginia, reported the Associated Press on April 21, 2009. Imagine if big brother could watch anything someone looked at, such as intimate moments with a loved one, or time spent with ones children. What if clips of such moments could be put on the internet, made to look like one had a hidden videocamera, when in fact the camera was in a microchip implanted in that person's brain? While this sounds like science fiction, the fact is the technology is here, but the regulation is not.

In a study published in the journal Nature in 2002, scientists rigged up five rats with miniature videocameras and devices that stimulated portions of the rodent brains to use them to move left or right, getting video of everything that they looked at.

In the 1960s, Tulane University implanted electrodes into people. BrainGate has a product, first tried in Matt Nagle in 2004 that detects brainwaves and turns them into motorized action. This is used by paraplegics to operate motorized wheelchairs, and other devices.

In 1999, the BBC reported that researchers had implanted a microchip into a cat that detected what the cat was seeing, and broadcast that into a computer, using the output of 177 brain cells. Researchers in Hiroshima, Japan are developing miniature cameras to be implanted in blind people.

Furthermore, the devices themselves greatly increase the likelihood of cancer, both through the microchips themselves, and the need for wireless communication coming from the most sensitive part of the body to such electromagnetic fields. A series of scientific studies from 1996 to 2006 found that the rate of cancer was between 1% and 10%, higher for mice and rats implanted with the microchip transponders for sarcomas, fibrosarcomas and other invasive cancers surrounding or attached to the implanted devices, according to a study published by S. Le Calvez et. Al. in Experimental and Toxicological Pathology in 2006.

"From a medical standpoint, obviously you worry about radiation with any electronic device," said Dr. Arun Patel, a general physician in Los Angeles (CNN, 2004).

With the critical need of the Fourth Estate: the media, to watchdog government, and uncover violations of liberties and abuses by the government, such devices could preempt efforts to uncover corruption in government, planned overthrows of the democratic process, and the start of World War III. Why is the media failing to cover this major scandal.

With the horrific abuses of civil liberties granted by the Patriot Act, the Department of Homeland Security is implanting people with microchips against their will. Though in gross violation of the Constitution, the secrecy granted by the Patriot Act is used to try to keep clandestine uses of such technology hidden. With absolute power granted by such secrecy, the very survival of the union – and our planet – is at stake.

Such devices could be put in the president while he slept, or in top military commanders. Access codes to nuclear weapons, and online access to drones and other military weapons could be gained through the abuse of such technology, as a means of causing World War III. Clearly we need more oversight into exactly what the Defense Department and others are doing with this technology.


**Chapter 7: Speaking Out, Working With the Courts**

I continued to speak on Viewpoint nearly every day. Viewpoint is a bastion of freedom for southeast Ohio. It is an hour-long program on 770 AM, and channel 5 on Nelsonville TV Cable. It has recently been expanded to 103.9 FM. They let anyone call in and speak for up to 4 minutes.

I filed a no contact order with Sasha Sigetic and Ross Martin. Judge MIchael Ward subpoenaed Sasha into court. I showed him my x-rays, pointing to the implant. He agreed to the order. She immediately

violated the no contact order with the brain implant.

At the Farmacy, Sasha said that she wanted to leave my house.

She offered me baby chickens. I was relieved, and immediately put an ad

in the paper for new tenants.

I purposefully got a parking ticket and took it to court. I

subpoenaed Sasha and Ross, going personally to Casa Nueva, and

handing it to Ross. I gave one to Sasha in the Farmacy. In court, I moved

to separate the witnesses. Ross went outside. They were still in violation

of the order to separate them because of the brain implant

communications. Sasha raised her hand and the persistent liar pledge to

tell the truth -- something as foreign to her as a distant galaxy.

I asked if they had any knowedge of the parking ticket and they

said they did not.

I also subpoenaed Jeff Benseler. He brought attorney Kennneth

Ryan who gave me an order to cease and desist ever communicating with

the radiology department at O'Blenness. Ryan also kept Jeff Benseler

from testifying.

The fact that they had to use a lawyer to keep a doctor from testifying over a parking ticket is such an admission that they have something to hide, I used it as another smoking gun that the media should cover. It was a clear admission in open court, or why would they do that? As usual, the media snoozed.

Both Sasha and Ross testified that they did not know about brain implants. I asked that they be given perjury charges. The judge ruled against me: I would get used to the unjust ways of Judge William Grim.

I was stunned at the horrific way that I was being treated. I soon learned I would be gang stalked and assaulted on a daily basis with an escalating indifference to the safeguards against violations of my civil liberties that were supposedly protected by the US Constitution.

An Athens County Sheriff deputy said "The Constitution is for Criminals." That is the mindset of those who have sworn their oath to protect and uphold the Constitution.

## Chapter 8: MRI at Holziers Hospital

I got an MRI of my head at Holziers Medical Center in Athens. When they gave me that shot in my right arm for contrast, they ripped my vein, causing permanent damage. Inside the machine, I was claustrophobic, as I thought I would be. It made a lot of noise. I tried my best to get through it. Afterward I went to the Athens farmers market, then home.

Holziers lied, a common practice among American doctors and said I did not have any implants or skull fractures. Using the same MRI data, doctors in Cuba would later find both skull fractures and a brain implant. When doctors lie to patients about the health of the patient the loss of the most fundamental liberty known to humanity is shattered. It is apparently done with National Security Letters, better called Nazi Spying Legacies.

Shortly before they implanted me, I set up an hour long meeting with two FBI agents at their Columbus Road office in Athens. I said that people were clearly breaking into my first floor area, and that kept me from leaving. I asked if they did any searching. They said that they did it with national security letters.

To force doctors to lie about a patient's own health is the most sacrilegious of endeavors by terrorists within a government too many Americans deem free. It is not free but the truth about the loss of freedom is hidden from view of the public by national security letters.

I stepped up pressure to get local media to cover this story. WOUB has a content officer. He like nearly all news media seem oblivious to the biggest news story on the planet: how the US government is putting brain implants into people against their will.

I was worried about getting brain cancer and saving my country from the terrorist attack by our own government. How many more people are victims, and how has their lives been affected. Brain implants are the most blatant violation of the 4th Amendment known to humanity. There is not a more egregious violation of privacy known than to read people's minds.

Still, there was little effect in my efforts to get the media to cover this story. I decided to start a hunger strike, and put out a press release that I would continue until the government admitted that they had put the brain implant into me.

## Chapter 9: Hunger Strike for Brain Implant Attention

I had done numerous fasts but never a hunger strike. I had fasted 11 days in 1995 protesting the attempted dismantling of 25 years of environmental progress by Newt Gingrich. My first fast was during the 1992 Earth Summit for a week, where I joined many others at Lafayette Park in front of the White House.

The 1992 fast was traumatic because it reminded me of starving in the Arctic Wilderness the prior summer. This fast was not that difficult, and I eventually started eating: giving up on the hunger strike after no media seemed interested.

My Mom called the sheriffs, and an officer came out. I said I was fine, and told them not to come back, but do admit that they had a brain implant in me. Chris Henry came out a few days later. I said he was trespassing and told him to leave.

Then an army of police came, and they kicked in my door, and put me in handcuffs. They took me to Obleness Hospital, where they forced me to give blood and urine against my will. What a massive violation of my 4th Amendment rights.

I was then transferred to the Appalachian Behavioral Healthcare: my first time in a mental institution and I wanted freedom right away. I could not believe that they thought that they had the right to hold me against my will.

I had already been eating for a week, and I had had a veggie burger cooking at the time Chris Henry was at my door the first time. I had recently started a hostel on the second floor of my house, and I did not know who would take care of it. I was in shock as my liberties were flushed down the drain.

## Chapter 10: Locked Up for Telling the Truth

In ABH, I was asked to sign a lot of papers, which I refused. I said that I wanted to go home, that I had been eating, and did not need to be there.

I was given a private room on Unit One North. I was in utter disbelief at the injustice that I was a victim of. Not only was I a victim of brain surgery and assault done against my will, and being locked up against my will, I was now incarcerated for no reason.

How atrocious, how evil our government has become: thugs working to mind control the very best and most patriotic of citizens.

I had nothing to do, and I could not get out. I spent the days preparing for trial, and cutting out clippings from the NY times. Criminals like Pat Kelly tried to make the mental health center my home. "That is where he belongs," The Athens Messenger quoted Kelly as saying.

In fact I had created one of the most sustainable homes in America, and 2 successful businesses, all on poverty wages, earning less than half the poverty line.

\*

4592 Bessemer Road; Nelsonville Ohio 45764. What a paradise I have created out of a very run-down place.

When I bought the property, there was trash all over, the walls were peeling off, and the roof needed replaced. Through thousands of hours of hard work, and lots of hard earned money from my speaking tours, book sales and renting the upstairs -- what most people would have called the nicest part of the house -- I created the paradise on Bessemer Road.

I love my house -- it is a Mecca of environmental and ecological right-living and a paradise for me, my hostel guests and wildlife.


## Chapter 11: Hand Dug Ponds Make My Property a Paradise

By creating three pond areas, one with lots of small ponds that occasionally turn into one large one when it floods, and 2 smaller ones, wildlife always know they will find water. With my fruit trees, wildflowers and "weeds," wildlife know they can also find food.

Because of this, there is always wildlife of great diversity on my property. Owls, woodpeckers, hawks, buzzards, doves warblers, cardinals, robins, sparrows, swifts, purple martins and scores of different species of birds, raccoons, squirrels, coyotes, rabbits, deer, fox, skunks, chipmunks, wood chucks and many more mammals as well, turtles, frogs, snakes, butterflies and more abound on my lovely piece of ground.

I watched a deer on my video surveillance at about 2:30 a.m. walking along my walkway just like a human, then toward where I have a pool of water for starting plants. Once I saw a deer live on my front steps while I was talking on Viewpoint.

In looking at my set-up the next day, I realized I had created the perfect deer-watering hole. The deer can get to water very easily, and have a perfect, comfortable and secluded setting to drink without even having to bend their necks down to drink.

I also witnessed a little blue heron enjoying my new large pond, along with Robert Anderson and his wife who were guests in my hotel. But it is unfortunate that so many people with mal-intent have come to the hostel. Many come because they don't like what I have on the internet, and try to break in to hack my sites.

## Chapter 12: Security to Counter Police Dereliction of Duty

Some may call me a sitting duck, but they don't know my security. I have designed my personal habitation with a spiral entrance full of security aparati. With three locked doors, video surveillance, and my large dog, as well as lots of hidden video cameras (only in areas where only I am allowed) I caught William Black robbing me with socks on his hands as you can see at www.repealthepatriotact.org.

With corrupt police often being the ones who break in, who can I call when break-ins occur? Some seem to think I don't like my place or my hostel because of all the break-ins. That is not true. What I don't like is the thugs who have gotten away with robbing me and assaulting me.

Those who allow the break-ins to go unpunished are committing a form of assault -- a psychological as well as physical assault. That is the reason I stay at home when I would rather be out fishing, hiking, going to coffee shops, enjoying the thousands of acres of forest in my backyard, etc. That is the reason I have lost a lot of faith in the government.

Most communities support authors. We are the ones who bring in tourist dollars. We are the ones who immortalize our time and place. I got assaulted by a brain implant, vandalized, robbed and assaulted by punches that broke my bones as well.

I had wanted to buy a house for a very long time in the area. I had put $1,000 down for a house near Lake Hope, not far from Athens, in 1997. Then, I was too ethical to keep my Ohio University job working to protect Dysart Woods.

I spoke out against OU's efforts to cut the buffer zone for Dysart Woods from 4,170 acres to a much smaller area, and founded Dysart Defenders. Instead of using the $1,000 to buy the house, I used it to protect Dysart Woods.

Since then, there had not been any real hope of buying a house. I had been part-owner of the Edges Community, then moved to a house on Peach Ridge, rented until an owner could buy it (for one room in a 3 bedroom house).

For several years I have lived without even running water at the New Convenant fellowship so I could focus on making my first real film, Caribou People. Other years, I have been homeless, having only a storage shed, where I set up a bed.

Local state forester, Ann Bonner let me stay in one of her houses doing work-trade for many months. I planted a lot of asparagus. Then David Walker broke down my door and began throwing 5 gallon buckets of water into my room of valuables. He finally got arrested, but many of my valuables were permanently damaged.

After that, I saw an ad in the paper, and moved to 222 Kontner St. in Nelsonville, 3 blocks from the public square, and the Mine Tavern.

I loved the community, and all the forest around it. Then I was assaulted on May 13, 2003. I was hit in the head with a brick and robbed. Crime Stoppers put out a $2,000 reward, but the assailants were never found.

That is where I became an author, publishing Arctic Quest: Odyssey Through a Threatened Wilderness Area about my 700-mile journey by foot and raft throughout the Arctic and Arctic National Wildlife Refuge.

I escalated my public speaking career with that book. I would hike up, by the cross overlooking Nelsonville, past the 911 emergency radio station and into the blessed Wayne National Forest.

After Kontner St., I moved back to Art Gish's place, where I had stayed for more than 6 months in 1994, developing my Christian roots far deeper at the Gish's. I learned film-making, and read my Arctic Quest book into an audio book.

## Chapter 13: Buying My First House

After getting a $60,000 settlement from the DC police because they broke 3 ribs while I was sitting in a demonstration, I began the exhaustive search to buy my house in just the right spot.

First, I looked into buying a four apartment house on West Washington Street in Athens. I knew that I wanted a duplex at least, because it would help to generate revenue, have people on my property to help watch and guard my property. It would also make for a more fun and social home than being by myself.

But try as I could, I could not get a loan to buy the house in Athens. Also, I really wanted a house that was situated next to a large expanse of public lands, preferably near a lake or river or the Wayne National Forest. For months, I looked at homes, and searched the internet and newspapers for listings.

Shane Montgomery told me of a duplex he had that sounded perfect, being next to a large expanse of national forest. The first time I went to search for the house I could not find it. I called and got directions again.

When I saw the run-down house, I loved it at first sight. I always wanted to live next to a forest, where I could just go hiking anytime I want. Now, I had my chance. And the price was right. I wish I would have done some more serious investigation and negotiating.

But I just asked for his absolute lowest price, and took it. That was $36,000 with $13,000 to be paid over 5 years with 5% interest.

Now the work began.

First, I went on a month-long speaking tour. When I got back, I pulled out the moldy carpets, and threw them away at great work (I had to cut them and roll them).

When Shane asked what he wanted me to throw out, I said to leave everything. I said that I could probably use most of it, and that I would recycle or throw out what I wanted to get rid of. I sure was glad later, after I learned masonry, that I did that. Many items like scrap metal would be used to prop up the old garage that was being pushed over by the hillside upon I lived.

Other items I sold for pennies for scrap metal that I wished I had saved. Even rusty metal worked fine in masonry, where the rough surface adheres to the mortar, or saved by the application of oil or paint. I used them inside of supporting walls.

The peeling plaster walls of my house I saved with joint compound by the 5-gallon bucket. In the bathroom, I learned masonry, pulling tile from the wall and putting it on the floor. The masonry skills that I learned would prove a value exponentially greater than the work. I learned to use the broken pieces of tile, all of which I would use in mosaics that are far easier and I think look a lot nicer, more artistic.

I bought a tile cutting tool, and a masonry cutting circular saw blade. I studied how to do the masonry as most tile jobs use, creating a grid, then making custom fit pieces for the edges.

Now, I don't do that form of tile, prefering to create art with my masony. It is also faster and easier. I use my knowledge of art, and how colors go together, to create art while putting floors on. I may spend more time overall placing tile in the place they will go, but it is pleasant creative time. It is time working creatively making a mosaic while enjoying a cup of coffee or beer, rather than what feels like unpleasant work time, as cutting tile feels like.

Also, I use broken tile pieces that most people throw away. That is not a small benefit. Using waste products -- even broken ceramic dishes and pieces of glass - is an example that can help enormously to save the planet. It also gives added use to the resources gained as we mine landfills in the future.

Once I learned masonry, so much more became possible. Suddenly, I could repair my crumbling concrete driveway and walkway with ease. I created a nice tile floor mosaic in my first floor area. I used re-used tile from Re-Use Industries for the masonry work that I spent hundreds if not thousands of hours to create.

I really liked the look, and I was a convert to "free masonry."

With extra mortar, I fixed my crumbling walkway. I also managed to fix my driveway by pouring the waste water with masonry extras (watered down concrete), onto it, then letting it dry.

## Chapter 14: Mental Hospitals Cause Psychological Disorders

Now, as I write this, I am locked in a mental institution, worried about what has happened at my house. I do know someone broke into my primary persona habitation. Otherwise my dog would not have gotten to the outside door as I was told that he did.

To just sit and worry about my prized possessions that I have safeguarded so well is the ultimate of torture. I can only hope and pray that I will return to a house undisturbed. My files and writings are so valuable, accentuated by a court order to file motions in my federal lawsuit against the very place where I was now at.

Meanwhile, a dream of building a pond that goes right into a new house where I can just lift a trap door and go fishing was born. What a dream life I would have, working on writing and film making while fishing at the same time.

I have already scoped out the pond, which would connect with my existing pond, and be quite sizable -- enough to grow lots of fish. I could also catch smaller fish then grow them larger in my pond.

## Chapter 15: The Interview With Jeff Stibel

In order to try to explain to the world that brain implant technology exists, I tried to set up an interview with Jeff Stibel, Chair of BrainGate. Braingate makes brain implants, like its wireless braingate2. His secretary asked me to send some of my books. I mailed a copy of my first book, Arctic Quest: Odyssey Through a Threatened Wilderness Area, and my second, Arctic Melting: How Climate Change is Destroying One of the World's Largest Wilderness Areas.

She called back about a week later and we set up a half hour interview by phone. I got permission to record it.

Brain implants are a powerful new technology that actually allows people to communicate using their thoughts, and for people to see what people are seeing, hear what they are hearing and even to place voices or other people's thoughts into people's heads. They have been around for decades, but have advanced enormously in recent years. Now, however, they are being misused by the Department of Homeland Security, Central Intelligence Agency and others to stop efforts to repeal the Patriot Act and to engage in torture against US citizens.

Paraplegic Matt Naegle was given a brain implant in 2004, which he used to surf the internet. DARPA and the Department of Homeland Security have spent billions of dollars on this technology, raising questions of its more nefarious potentials. In Ohio, two people have reported that brain implants were put in them against their will: I know, I am one of them. With the advancement of this technology, the thought police are real.

Jeff Stibel is a world-renown expert in brain implants, as the Chairman of Braingate, a company that recently unveiled their Braingate2, in which people can communicate wirelessly with computers just using their thoughts, through what is called a brain-computer interface. I interviewed Stibel by phone, with the audio clips of the interview posted at www.brainimplant.info. Here is the transcribed interview:

Chad Kister: What's your background on brain implants?

Jeff Stibel: I was in a doctoral program at Brown University in brain science. And I have helped to start a handful of companies in the field.

Chad Kister: Is the technology there to read people's thoughts?

Jeff Stibel: The technology that I am associated with, I'm the chairman of a company called Braingate, does just that. That's exactly what it does. This is a microchip that is implanted in someone's brain, and we effectively read the thoughts of the person. In our case, what we are trying to do is to control movement. We can interpret that people are trying to move a mouse curser, and then have that mouse curser move, just with their thoughts. You can play videogames on a computer using this technology.

Chad Kister: What is the potential of the brain-computer interface?

Jeff Stibel: The potential of Braingate is absolutely tremendous. For people who are suffering through injury, trauma or otherwise to have the ability to both help and allow them to communicate, but also to help them enable prosthetics, wheelchairs and other devices just through thoughts. It is absolutely tremendous for the handicapped population.

For more fully functional people, the ability to use your thoughts to control electrical devices is really the transformation of science fiction into reality.

Chad Kister: What is the potential to see what people see, like a video-camera, either with a miniature video-camera or through the neurons? Have you heard of that potential?

Jeff Stibel: I have. Again, like everything that is being processed through the brain, vision is processed through electrical and chemical reactions. The ability to actually tap into those electrical connections, similar to what we do through Braingate, is real. That can be used for good or evil, for commercial gain or to help humanity. But the potential is there, it is tremendous.

Chad Kister: What about the ability to hear what people hear?

Jeff Stibel: The same thing exactly. You are talking about electrical impulses. The ability is absolutely real. You can use those for many different purposes. But that is certainly something that is not out of the realm of reality. There is one that is working to make the blind be able to see. You effectively have a brain implant that hooks to what looks like sunglasses. The glasses are actually processing what you normally would be looking at except this person is blind, and then feeds that information through a computer chip directly into the mind, to give that person the sensation that they are actually seeing something, and it works reasonably well.

Can you read people's thoughts in words? Say someone thinks "I am going to go to the store." Would there be a way to hear that person think that they are going to do that?

Jeff Stibel: Yes, absolutely, whether you are thinking in words, or symbols or otherwise, in the brain that is all translated into neural activity. You can tap into that neural activity, and uncode it, then you can determine exactly what that thought is.

Chad Kister: What can you decode at this point?

Jeff Stibel: Right now we are still in the very early stages of this technology. And again our technology is being used principally to help paraplegics, locked-in patients gain mobility, and be able to communicate with others. We are not trying to look into their minds to see what they are thinking, we are looking into their minds and help them function better: to help them communicate, help then to move around in their environment.

Chad Kister: What about with Alzheimer's patients?

Jeff Stibel: The potential is there, but a lot less so. Alzheimer is a degenerative disease, so the brain is deteriorating, just as it is with mad cow disease. That would not be a particularly good application for a Braingate technology. Other brain implants, however, are working to help with Alzheimer patients.

Chad Kister: Have you been able to use them so that people with prosthetic arms are able to move them with their thoughts?

Jeff Stibel: We have done research in that realm, and there is research at Duke, MIT, CalTech and Brown right now in animals trials, not yet in human trials. We believe we will be able to get there, and we will be able to control a human prosthetic arm.

Chad Kister: What about with uberveillance? Are you familiar with the term uberveillance?

Jeff Stibel: I am not, No

Chad Kister: It is about being able to monitor people using a brain implant device. The Defense Department and the Department of Homeland Security have much more advanced brain implants, and appear to be using them to engage in thought-policing, according to two Ohio victims, both of which were documented to have brain implants. Imagine what it would be like to have people monitor your every thoughts, your every move, even your dreams.

Chad Kister: How are they powered?

Jeff Stibel: It is powered in the same way as any other electrical device is powered, with a battery. It is literally a silicon-based computer chip that ties in either through wires or wirelessly into a computer.

Chad Kister: If it's wirelessly, how would you keep it charged?

Jeff Stibel: There will be a battery in the small system that is in the brain, or just above the surface of the cranium, the skull. And that battery will transmit that information to the computer, which is with a larger battery of course. It needs very, very little power. You can also go and do some research if you go to braingate.com. There is a good amount of information there.

Chad Kister: What about the ability to use electrical power from a person's neurons, or from a fuel cell, so it would last a long time?

Jeff Stibel: That's possible. The way the brain generates power and gets its energy is through oxygen and the blood supply. I think we're a ways away from being able to tap into that. But interesting idea, very interesting idea.

Chad Kister: If it is done wirelessly, how long would the battery last, and how would you charge it?

Jeff Stibel: Years and years and years and years.

Chad Kister: Because it just uses so little power?

Jeff Stibel: You are talking about an implant the size of a small M and M. It is tiny and it uses very little power.

Chad Kister: How far can the wireless frequencies go?

Jeff Stibel: Unknown right now. With the technology that's out there these days, with Bluetooth or Rf frequencies, it should be a reasonable amount.

Chad Kister: Have you heard about law enforcement with an interest in being able to interrogate people or question someone with their thoughts?

Jeff Stibel: I have.

Chad Kister: Is it the Department of Homeland Security, or which agency have you heard interest from?

Jeff Stibel: I think there's absolutely interest: there has always been interest. But that is a long ways away. But a lot of the funding in the academic lab comes from government agencies, DARPA, the VA. It is no surprise that there is more broad interest and applications in the defense side and the law enforcement side.

The author is in the process of documenting two cases in which citizens have been interrogated with brain implants against their will. The Department of Homeland Security apparently thinks it can use the Patriot Act to engage in thought policing. This is the most egregious violation of civil liberties on the planet. Imagine not being able to escape. How would one exercise his or her Fifth Amendment rights to be silent? Imagine 24-7 interrogation, with no place to escape.

This is happening, and the perpetrators, who appear to be police, are betting that people will not understand this technology, so they will get away with it. One of the individuals said that they are trying to send him to a psychiatric ward, because he has been so effective in mobilizing concern for civil liberties protection in the community. He has also been on the forefront in the effort to repeal the Patriot Act. We need greater oversight over the use and potential misuse of brain implants. Above all, they must never be used against someone's will, as was done in two instances in Ohio.

Chad Kister: How hard is it to transfer the brain's code into a computer interface?

Jeff Stibel: It is incredibly difficult. It's taken years and years and years of technology advancement to be able to do it. You are basically converting a biological system into a computer system. But fundamentally, the brain computer system -- our neurons -- isn't that different than a computer system. So fundamentally, in theory, it is fairly straightforward. But in practice it has turned out to be incredibly difficult.

Chad Kister: Do you have to go to different portions of a person's brain to get someone's thoughts, or what someone sees, or what someone hears?

Jeff Stibel: Neuronal mapping has been around for the better part of 100 years, and they are getting better and better and better. Depending on what you are trying to do, you have to put the computer chip in the right spot to be able to read the right type of neurons. The brain is always active. You want to get as close to the motor-active, the hyperactive neurons as you can. If you are trying to analyze, leverage or interpret motion in the brain, you need to put the electrodes as close to the neurons that are interpreting motion as possible. The same is true for speech, vision or anything else in the brain.

Chad Kister: Where would you put an implant to monitor someone's vision:

Jeff Stibel: The vision part. It's called V-1: It's a part of the cortex.

Chad Kister: What about to monitor thoughts:

Jeff Stibel: Everything that's happening is someone's thoughts.

It depends on what you're talking about. The pre-frontal cortex which is closer to the front part of your brain is generally someone's decision-making center.

Chad Kister: What about for hearing?

Jeff Stibel: Again, this all comes down to where the neurons are more active in the brain, that is where you put the implant. Where the neurons light up for hearing is where you want that chip to go.

Chad Kister: How many neurons do you need to connect to make a Braingate function?

Jeff Stibel: Surprisingly few. It can be as few as a couple dozen up to; they've measured a few thousands, and it did not seem to make much difference. The level of complexity is much greater in terms of implanting the system (when more neurons are connected). They did it with monkeys, versus what we did with Braingate, which is a single chip. But you need to activate just a few neurons. Again because they light-up in a network effect, they're all associated, so you just need to tap into the overall connection and activity in the right spot. You don't need to get them all.

Chad Kister: How long does it take to put one in someone?

Jeff Stibel: It is reasonably quick, the surgery is done within a single day.

Chad Kister: Does it take an atomic microscope, or how would you actually see to attach a neuron?

Jeff Stibel: You are really looking for the placement (of the implant). It is brain surgery. You lift the skull, you go to the spot (and push the spikes attached to the implant into the brain). Again, I am not a neural surgeon. It is a lot more complicated in practice. Other than the fact that it is brain surgery, it is not a highly complex surgery.

Chad Kister: How are the neurons attached to the wires?

Jeff Stibel: Neurons are absolutely tiny. The placement basically: you are looking at an electrode, a microchip the size of a paper eraser and on them a hundred spikes. There are a hundred spikes is the best way to describe them. You push those spikes into the brain. There are a hundred billion neurons in the brain, with a hundred trillion connections. It is hard to miss them. The spikes are designed to feed into those neurons. And then from there either you have a wireless connection, as in Braingate 2, which is what we just got FDA approved. Or in an earlier version -- you may have seen it on 60 Minutes when we were featured -- the 60 Minutes version was a system that literally connected your brain to the computer with some fairly heavy elaborate wiring.

Chad Kister: What about the potential to advance people such as a politician in a debate?

Jeff Stibel: It is very much science fiction at this point. That said, this is something that Sergey Brin and Larry Page have been fond of for years: these are the founders of Google. There is the ability to connect all of the world's information to the brain through the internet. So if you have a Braingate2, and you can connect to the internet, you can gain access to all of that information.

Chad Kister: How many years off do you think that is?

Jeff Stibel: In limited instances, you can do that now. They can log on to a browser, and surf with a limited capacity. But in terms of doing this for the masses, you are talking about years and years. Because it has to be safe, it has to be bulletproof. And it has to be simple. This does involve brain surgery.

Chad Kister: What are the dangers of brain surgery and putting in a brain implant?

Jeff Stibel: With brain surgery, if something happens to your brain, you can be in serious trouble.

Chad Kister: What about some of the potentially more nefarious possibilities, such as if someone were to put this in someone without their consent?

Jeff Stibel: It is always a possibility or a risk as with any new technologies. When they invented the wheel, as good of an invention that that was, there was the risk that it could do harm, to be used for tanks and to wheel in cannons. There isn't a technology out there that can't be used for good or for evil. It depends on the intentions of both the inventors and the people who use them. There is far too much good that this technology can offer to think about or concern yourself with the potential harm.

Having found two people so far that have been implanted with brain implants against their will, along with the interest among law enforcement agencies to thought police means that people need to be educated that this technology exists. The cart-blanche powers of the Patriot Act appears to have allowed for police to place these into people against their will, in a gross violation of the U.S. Constitution, which trumps the Patriot Act.

The inherent secrecy of the Patriot Act, with National Security Letters used to silence any potential leak of information, can keep the thought police safe from public scrutiny. Safeguards, such as methods to detect brain implants are critical. With the use of these to try to make people look like they are insane, to discredit and lock them up, we must concern ourselves with the implication that our thoughts are no longer sacred, unless we reign in the misuse of this technology.

First I uploaded the audio to www.brainimplant.info

I noticed years later that the audio was not working, and had to re-upload it. Thank God I was able to find the file. I live in extreme poverty because they constantly damage electronics, and threaten to lock me up.

## Chapter 16: to Canada for an X-Ray

I decided to go to Canada to get an x-ray. I drove up North into Michigan and went into Canada across the bridge from Detroit, into Windsor. I found a hospital and went to the Emergency Room. They had me in Tri-Age. I explained that the US government had put a brain implant into me, and I tried to show them the x-ray on my laptop. They had me fill out forms and wait.

I demanded an x-ray from the side of the head to prove it in Canada. They put me in pediatrics, and treated me like I was an infant, and scolded me that my laptop was not clean. I said it was clean but people break in and cause it to be filthy with their filthy hands. They seemed to care nothing about the obvious brain implant in my x-rays. I wondered how big the conspiracy was.

Canada tried to charge me more than $1,500 for one x-ray, and did not give an accurate diagnosis. They are clearly just another pawn of the US imperialist brain implant agenda. We need to admit to the use of this technology, if we are to be a free, transparent government. Otherwise, as we do now, we live in tyranny, where even the freedom to think a free thought is no longer sacred.

I stayed in a cheap motel a few miles from the hospital, then went back the next day to get the images which never loaded on my computer. They would then harass me and try to collect on more than $1,500 for them to poison me with radiation and just tell lies. They could have just used the x-rays from Obleness Hospital in Athens.

I began to ponder how in the world I could ever get justice in what appeared to be an entirely unjust world. I thought for sure once Canada realized what the US was doing, they would immediately take action. But I should have realized that because the US is the dominant power, with veto power on the UN Security Council, that there is nearly no way to challenge their injustice. Canada is likely implanting people as well.

I decided to try Cuba. Because they were not aligned with the US in geopolitical politics, I thought that maybe they would tell the truth, and help our country's future by finally exposing America's biggest scandal: brain implants done against people's will.

I researched the laws, and found that it would be legal to travel to Cuba, as long as the purpose was other than travel. My purpose is never to just travel, no matter where I go. I always write for future books and other writings, get video and photos for films, and use all travel to increase my scholarly knowledge, science of areas ecology, etc. I also give presentations about areas, using my experiences and multi-media from such places.

## Chapter 17: To Canada to get to Cuba

I have always wanted to travel free, without a necessary destination, itinerary or means of movement: just go as the opportunities present themselves, but always for a deeper purpose. Never have I had a more important purpose than this: to expose the brain implant terrorism that I am a victim of. The US government has put a number of brain implants into me to make me a targeted individual, among the most tortured and harassed as could be imagined. But the torture is invisible.

What a sadistic, evil government that would spend so many resources just to harass citizens, through a Cointelpro (FBI) or Psyop (CIA) type program intent only on doing harm. These terrorists need brought to justice, with new forms of punishment needed to deal with this new from of terrorism, torture and harassment. We could long have had all of our energy through solar and wind had the resources that went into this gone toward good, rather than evil.

Mark Twain said, "loyalty to the country always, loyalty to the government when it deserves it."

That I need to travel to Cuba to get a proper diagnosis is a sad day in American history. The Patriot Act has gone to such evil extremes as to force doctors to give false diagnosis, a sure path to murder. Whatever happened to the hippocratic oath?

I gave plenty of warning to the powers that be. Lenny Eliason said that he had a document from 2009 about a leak, saying that is all I should need. But I was not able to get it because I don't own a car, so it was mailed.

I packed well, and light (relatively), keeping a detailed list of everything packed so as not to miss anything important. I packed my dry suit, raft and paddles, which I planned to use for fun and to make sure I got into Canada, no matter what the border guards attempt. I could not imagine they would do anything to stop me, as I would make a very big issue out of it, and get in nonetheless, blaming them for the dangers they forced me to undergo.

Skipped were the dress clothes: they would be too hard to keep pressed and unstained. I will have to hope and pray that the humanity in people will see through my lack of suit and tie appearance.

I woke up and announced on the radio the protest against the FBI and their assault of nonviolent peace activists in Columbus. Forty-nine rallies were scheduled around the United States.

I went up and knocked on Amber Hager's door, a tenant, to see if she could give me a lift to Logan, where I would board my dog, and take the bus to Columbus. She did not open the door. I called Westside Taxi, and they said they would have a driver right away. She said, "it sounds like Chad." She said that the driver knew where I lived. I finished packing in a hurry, turned on all of my surveillance equipment, and went outside just as the cab was pulling up past my house. I put my dog Dude on a leash, and went down the driveway with my two packs, and put them in the cab.

I explained to the driver how I was starting a hostel and a kennel. The kennel could use a building that was under construction and plenty ready for pets, but not finished for people to live in. And, putting hostel guests in a place where they could access pets is a great way to create a symbiotic relationship that benefits both the lonely travelers and the needy pets. "I enjoy hanging around cats and dogs, so I enjoy it," I said, "and it helps to pay the bills." That building also has an EPA certified wood stove with glass windows. That should help create a very quality atmosphere, like hanging around a campfire.

I called two kennels. The first one, being 50 cents cheaper, was booked up. The second did have room. The cab pulled up, and I brought my dog in. I had already told the woman that I was in a cab on my way. She just asked about shots. I showed her my ID just in case she needed it, and that was it, I could pay when I returned. I was pleased at how fast I checked my dog in, because the cab driver was waiting.

I next went to the Old Dutch restaurant, and purchased a ticket, and a cookie, and waited for the bus. When the bus arrived, I hurried to get out, but the bus did not stop. Instead it began to pull away. I ran after the bus, and the driver eventually stopped. I gave him my ticket and he said that I had to wait outside when the bus arrived, because very few people got on the bus in Logan.

## Chapter 18: FBI protest in Columbus on the way to Canada

The bus had wi-fi, which makes traveling so much nicer, and more productive. I could not believe that more people did not take advantage of the $10 bus ride to Columbus.

The bus stopped at the Columbus airport, then went on in to the Greyhound station. There, I walked north, along High Street, looking for a Chinese buffet. I was quite hungry, and thought that would be a good way to save some money and fill my tummy.

On I walked, north on High Street. I walked through the Short North and very aesthetic brick streets, murals, and art shops. While restaurants abounded, none were all you can eat and I was very hungry, so on I walked. Soon I was at the Ohio State University campus, where I have been many times, my Mom got her PhD at and I have given a presentation on climate change at. I announced the upcoming rally at 5 p.m. at 15th and High in a loud booming voice whenever crowds appeared, such as at bus stops, and on campus. "The FBI has been attacking nonviolent activists. Come to a rally at 5 p.m. to protest the FBI" I announced.

I continued north, stopping in a few Chinese restaurants to see if they were buffets. Finally, several miles from the Greyhound station, I found an all you can eat Ethiopian buffet next to the Taj Mahal Indian restaurant. Shortly after 9-11, a crazy racist ran his car into that restaurant, showing the hatred and insanity rampant in America.

After 9-11, the American police state gained tremendous powers. They even put brain implants into whomever they want, and get away with murdering and beating whomever they like. I was beaten nearly to death by Washington DC metropolitan police, and no one was ever prosecuted (though I did win a civil lawsuit against them).

The same officer is still in uniform, undoubtedly assaulting more nonviolent citizens and activists like me, for no reason except to get their thrills of harming innocent people who they do not like. The absolute power granted to police has made many absolutely corrupt.

America is very similar in too many ways to Germany in the late 1930s and early 1940s, with a secret, Gestapo police that murder and attack in secret, with no one even knowing what is really happening. We have death panels of murderers paid by our taxpayer dollars (The NSA kill list and more), and police who torture and beat people for no reason whatsoever. We have the FBI who break into peoples homes, burglarize them, poison people and get their thrills by trumping up fake charges against nonviolent scholars and activists.

Doctors are allowing them to put brain implants into people against their will, or they are involved in covering it up. Just because they are forced to do so through National Security Letters, does not change the fact that they are complicit, and need to tell the truth.

I managed to leave my cell phone in the restaurant, having put a napkin on it, then walked out. I realized it quickly, and turned around. The restaurant owner said that the taxi driver who was in his restaurant went to find me. I had shown them my books, and told them I was on my way to Canada. I was pleased by the kindness of the restaurant owner: if only more people would be like him: helping people instead of assaulting people as the nefarious police state terrorists are doing.

I walked to 3rd and High, more than a mile I would guess, and had a beer. I then went to the rally, at 15th and high, where people held signs against the FBI. I handed out a lot of literature, saying "the FBI is locking up non-violent activists."

Some people took literature, and some did not. I explained to the people that one must never get worried about having one not take literature, but that one should always just keep at it, no matter what. Lots of people do not want to waste paper, I said. Many people give out literature for advertisements, so people learn not to take it. But once they know that it is for a cause, they tend to take it. Thus explaining the cause helps people choose to take literature.

They gave me their extra literature, and I handed it out, while walking toward the bus station. I stopped in Krogers on High Street, the same one that I visited numerous times when I worked on my Dad's apartments nearly two decades ago.

I bought some caffeine pills, multi-vitamins, and snacks. I wrote some on my laptop in the station, then boarded the bus to Cleveland.

## Chapter 19:  Greyhound Bus to Canada

In the station, someone leaned on my laptop, causing a great amount of damage.  There was now a handprint on the screen, but amazingly the Dell laptop still worked: it was just harder to use because of the hand print permanently etched into the screen.  I complained about my damaged laptop, saying it was clearly intentional.  They did nothing.

I bought an overpriced bacon cheeseburger, fries and Dr. Pepper, and had a refill.

On the bus, well after midnight, I caught a little sleep as we drove very fast through heavy snow, past semis that had slipped off the road.  There were nice views of Lake Erie.  In Buffalo, I caught the bus to Toronto, and soon we were in the customs line.

When I got to the customs official, I said I was an author, then showed him my book.  I said that I was there to write a magazine article about Farley Mowat, something I desperately want to do, though I was unable to message him on Facebook.  One of my publishers, Greg Bates with Common Courage Press gave me his email, and I figured I could find it on gmail and email him, or otherwise find a way to contact him.

He lives an hour away at Port Hope. I found out later that he lives in his Mother's house.

The customs official asked what I had. I said I had a raft, a dry suit and paddles. He let me through.

Someone left a bag on the bus, and someone else said that they were missing a bag, and the customs official put it up on a table not far from where I was. He found a prescription bottle. He removed it, and looked at the name.

Being the victim of numerous violent crimes, many of which are unsolved, and hundreds of property damage crimes, near all of which are unsolved, I find too many police to be thieves, looking to pocket money rather than helping to solve crimes that have victims. There are, luckily, exceptions.

I remember being on the Sky Train in Vancouver, in western Canada when a police officer walked through to check to make sure people had their ticket. I showed him mine. An Asian man next to me did not have a ticket. The police officer took his wallet, pulled out a wad of cash and threw the wallet back at the man. I was too afraid to say something then, but I filed a complaint with the police from my hostel's free phones later.

Back on the bus, we waited for people to finish getting through customs.

I slept most of the way to Toronto, despite the fact that it is an exceptionally scenic drive, one that I have done many times. There are great views of Lake Ontario.

I awoke as we pulled into Toronto, and I remarked that it is a very nice city. Street cars went around the city, running on electricity and rails, exponentially more efficient than automobiles or airplanes.

What a beautiful city Toronto is. At the bus station, I found a map, trying to recall where the numerous hostels that I had stayed at were. I went to a Starbucks, got some coffee, and sat down, opening up my laptop. I tried to get an internet connection, and there were numerous, more than a dozen, but most wanted to charge about ten dollars.

I befriended a German who I found out later worked for pyramid solar. He showed me which wireless was really free, and I found a hostel. I called, and found out that it was just a couple of blocks away.

I then chatted with the person, and showed him my x-rays. He suggested that I fly to Germany to get it diagnosed. I said that Cuba seemed less likely to be controlled by US Imperialism, and it would use a lot less Greenhouse gas emissions to get there, as well as having a more pleasant climate this time of year.

He gave me his card, and I sent him an email.

At the Dundas Hostel, I was greeted by a beautiful young blonde Canadian woman at the desk. I took off my $400 custom fit boots that were donated by the Mast General Store in North Carolina for my fourth expedition to the Arctic.

She asked if I wanted a 6-person or a 10-person dorm room. I asked for the 10-person one, at 20 dollars, being two dollars less than the other. I found out later, to my surprise, that she too lived in the room, next to my bunk, and was apparently staying work-trade. She had a boyfriend, who liked to camp in Montana and had a -40 degree sleeping bag.

I found that I was in complete paradise from a culinary perspective, with more sushi places than I had seen in Japan, Thai and being next to Chinatown. I walked over to the Ontario College of Arts, and was amazed by a building that was as unique and among the most beautiful I had seen, since the churches in Florence Italy when I was a child.

I cruised around, searching for an Asian food bargain, and stumbled into a library. As is my custom, I asked to see if they carried my books. I chatted for a long time with a librarian, who talked much about Farley Mowat, and Carsten Hauer, whom I had met in DC when lobbying for the Arctic Refuge, just as they had arrived from the airport. We went to a reception in the Canadian embassy, where I have been numerous times in lobbying visits, enjoying their fine wine and cheese.

He said that Farley was not staying at his Mother's house, and he suggested that I go to a canoe museum not far to the north from Port Hope. I noticed a restaurant with all you can eat sushi, and decided I just had to eat there, after I had worked up a sufficient appetite to justify the $12 expense. I had had a $10 Indian buffet the day before.

I stopped by the Toronto Police department, which was on Dundas, the next day. Two bicycle police were outside of their department, and I was pleased with the zero greenhouse gas emission method of policing. In the department, I spoke with a female plain closed officer, and showed her my x-rays. I said that the US government was violating Canadian laws.

She asked how I knew it was the US government, and I said that it must be them, as who else would do it while I was in Ohio.

The officer said that I should see a doctor, and get it out, and then they would have evidence. I said that I had had x-rays in Windsor. She said why had I not gone to a Windsor police department. I said that I had been flabbergasted that I could not get an accurate diagnosis, but that I should have gone to the Windsor police. I asked if they could test for wireless frequencies using a spectrum analyzer, but she said that she could not. I said that I did not want to have to fly to Cuba to save my life. She said, "Why would you want to go to Cuba."

I had to go, and already had a license, as it was a matter of life or death. Self defense and the necessity plea made it clear that there was no way that I could possibly violate any law. Also, I already had a license to go to Cuba from the State Department.

I walked over the Mount Sinai Hospital, and asked for Neurology. They said to go to Dr. Gordon, on the fifth floor. I went there, and two women said that I would not be able to see Dr. Gordon, effectively pushing me out of his office, and back toward the elevator. I said that the

US government was hypnotizing me to convince people to go to war with

Canada, and that they really needed to do something, all the while

showing them my x-rays, and explaining the implants, while showing

them my scars.

I said that it was a life or death issue, and that I just needed a

proper diagnosis.

They said that I should go to the West Toronto hospital. I said

OK. I went back to the hostel, then walked over to the West Toronto

hospital. I asked to see the neurologist. They sent me to the neurology

department, and I showed them my x-rays. They said that I would have

to be referred by a doctor. I had a referral by Nelsonville, Ohio

neurologist Dr. Getachu, but I did not have it with me. I decided then

and there that I would have to go to Cuba.

I got an all you can eat sushi, a major splurge, but with the

amount of sushi that I ate for $12 Canadian it was well worth it. I had

numerous pieces of sushi, and sushi pizza, trying to be very careful to

avoid endangered fish. I mostly ate salmon, tofu and avocado.

I had lots of green tea, then water.

## Chapter 20: Finding an Affordable Flight to Cuba

Back at the hostel, I watched CSI with people from all over the world. I said CSI should do a segment on brain implants. I said that I was going to Cuba. I packed up, and walked toward a bus stop where there was a shuttle bus going right to the airport, about a hundred feet away. I went out, searching for the right bus stop. I found a TV station, and gave them one of my x-rays. I then went to the shuttle bus, and got a sticker price at the $22 price.

I went back to the hostel, and found out that I could pay $3 for a ride on the subway (with far fewer greenhouse gas emissions, and a much more social and interesting journey). The subway was just a block away. I went there and took the subway, which was remarkably often, over to the stop to switch to the green line, going to the airport. I took it to the end of the line, at Kipling Station, where I got a transfer, and a free bus, the 192 bus, out to the airport.

I went to all of the airlines, standing in line for a great amount of time, then tried to get a ticket to Cuba. Canadian Airline was $849 to Havana. Another place was $714. I went back in line with West Jet, and they quoted $379 round trip (about). I exchanged money, then they said it was not available.

I went back and exchanged all the money back to American, including the exchange fee. It was enormously cheaper to change money at a bank.

I took the bus back to the subway back to the hostel. I got a gyro at the restaurant just downstairs from the hostel, along with a juice drink. I added some vodka to the juice drink, and worked on my laptop in the hostel.

I continued to encourage an Indian, who seemed very interested in my health and situation, to diagnose my x-rays and do something about it.

I walked with a group from the hostel over to a bar about 15 minutes away, then decided it was too noisy, and came back, still enjoying the experience, without the cost. I got some sushi and a juice drink for $8.

I checked on flights to Cuba via the internet in the hostel, and tried to book one, but for some reason my debit card would not work. The flight was for January 29, 2011, a Saturday. The Dundas hostel was booked for Friday. I called around, and found a room at the Global Backpackers Hostel. I then walked around, carrying all my gear, looking for organic coffee.

I had found one nearby, but could not locate it again. After about two miles of walking, I found organic coffee, and stopped in. I found out that my cell phone was set to forward calls, but my phone would not let me change it back. I called T-mobile, and they set it to receive calls.

I then found a nice Jamaican restaurant, advertising free internet. There was not free internet, but the food, barbeque chicken and orange rice, with a little dark green salad, was fantastic.

I went back, looking for a bank. I stopped by the Canadian Attorney Generals office, and showed them my x-rays, asking to file a complaint against the United States. They said that they could not help me, and looked angry. Very annoyed at their dereliction of duty (which I should hold accountable in international court), I told them that I guess I will have to go to Cuba.

I found a bank, and exchanged money into Canadian at a very reasonable price, though this was the first time I had been in Canada where the US dollar was weaker than the Canadian. The subway was in the same building, a mall, which is a wonderful way to encourage mass transit, because one does not have to wait out in the cold, sometimes rainy weather to catch the train.

I took the yellow line to the green line, then walked to the bus to the airport. I managed to get on the wrong bus, and got a good photo of a building covered with solar panels in the process. I took another bus back, their being surprisingly frequent, then caught the right bus to the airport.

I stood in the long line for West Jet, their being by far the best price, and bought a ticket to Cuba. I was asked if I wanted to share the information with anyone, and I said no.

With the ticket in hand, at $413 round trip, I was overjoyed, knowing that I had a real shot at exposing the major act of terrorism that the US government caused me. What an evil, sadistic government that would perform brain surgery against someone's will, then engage in thought torture, destroying my career and locking me in a mental institution.

## Chapter 21: Waiting for the Flight

I took the bus back to the subway, and checked the map to find the nearest subway stop to King Street and Spadina Avenue, near the Global Village hostel. I could have gotten a transfer and a free ride on the street car, something I have always wanted to ride on, but instead ended up walking several blocks. I paid for the quad room and dropped off my stuff.

I had walked past an outdoor store, and I went back, and bought some camp suds. The worker suggested that I talk to the owner about their carrying my books. I looked into getting some crampons to walk on ice for later Arctic expeditions. The black diamond seemed the best brand, being by far the lightest. But for now, I was going south.

I lay down, being successful in my endeavor to get a ticket to Cuba. I went down and did a load of laundry, then bought a pint of Amsterdam beer and enjoyed it in the bar. There was a pool table, and a pacman game, and many pleasant people from all over the world. Hostels are where you can really meet the world in one place, while saving an enormous amount of money. That is one of the many reasons why I started one myself.

Back upstairs, I set my alarm for early in the morning for my flight to Cuba. Two Australians came in, and we chatted some. I showed them my x-rays. They seemed to be familiar with Dr. Michael Michael, a scholar on brain implant terrorism, which he called uberveillance.

The bunk creaked at every movement, being metal. I was learning much for my own hostel. I decided then and there that I would not have metal bunks, but would make them of wood, as I planned anyway. I would use reused wood, of course, to save trees. I planned to get work-trade from low income people, to use empty beds to get work effectively for free. I already had done that, as most hostels like the one I was in did. This effectively lets people live without spending money, one of the greatest services one can do.

My hostel is an absolute paradise, with a hiking trail going into miles of trails in the Wayne National Forest. I hand-dug ponds and put in fish to eat mosquitoes. I planted lots of flowers and fruit trees with the freshest finest fruit one could ever have, free from chemicals and as fresh as can be.

I will never forget the taste of my first peach, picked this past fall, right off the tree. That taste of freshness is incredible: divine. The longer fruit is off the tree, the more the taste is lost. I grew up with a peach tree at my parents house.

I awakened early, and fumbled to turn my alarm off as fast as possible, knowing I was inadvertently awakening the Australians. I pulled all my stuff outside, then realized a good use, if not the purpose for the extra room that was well lit on each hall. I had much experience with hostel life, and how to get up early and pack without disturbing other guests.

The key is to get everything out of the room, then close the door and pack in an area where the noise from packing will not disturb other guests.

I packed up, then went downstairs. Fortunately, I found a hostel worker, and was able to get my key deposit back. When leaving in the early morning hours, sometimes one must get the deposit back the previous evening or night, as happened to me in San Diego in the early years of this millennia.

I walked out, looking for a nice organic coffee and local food shop. But everywhere was closed, and I ended up at a McDonalds, a place I try hard to avoid because of its size, and lack of organic food or beverages. I always like to give to the underdog with my scarce dollars, not the billionaires.

On I walked to the train station, wishing I had the extra money for a street car. I will surely get around to taking one at some time. I needed every dollar to get through this journey, knowing getting to Cuba would be expensive.

I got on the subway, which luckily was running at 6 a.m., and had the train car to myself, a first in Toronto.

I transferred, and got on the green line toward Kipling Station. I talked with people on the crowded subway, showing them my book, and my x-rays. I then got on the bus, and was in the airport. Thinking, erroneously, that I had plenty of time, I completely repacked my blue computer backpack, making sure that I did not have a knife. I discarded a few extra papers, and got in the very long Westjet line.

After quite a while, a westjet employee asked if anyone was going to Caya Coco. I said that I was, and he said that I did not have much time, but that I should be alright because I was near the front of the line.

I began to worry. But I checked on my big backpack, and went on to security. I went through security fine, then went on to my gate, with plenty of time. I picked up a Toronto Star, which turned out to be among the more interesting papers I have read in my life. I read about the turmoil in Egypt, and the effort to remove Mubarack. Luckily, Tunisias dictator had been removed.

## Chapter 22:  Aloft to Cuba

I got on the plane, and soon we were in the air, on the way to Cuba. I liked how the screen told me where we were, and how high we were. This helped to place the geography that I saw. I took many photos.

The landing was slightly rough, reminding me of Arctic landings, with a less than smooth runway. But it was plenty safe, and the whole plane cheered and thanked the pilot after we had slowed down. I thought this a little comical, and amusing, with the social psychological nature of a crowd being manipulated by the group. Still, everyone was safe and happy, and that is what really counted.

I wondered what the Cuban government would think about my not having a place to stay for the night. I went into their security area, and tried to communicate with a woman who knew mostly Spanish, showing her my passport.

An immigration guy came up and chatted a bit. They wanted to see my health insurance, which I showed them. I said that I was writing a book about Hemingway. Having recently listened to his book Islands in the Stream, published posthumously, I was very interested in seeing the bar that he hung out at, and the places where he fished.

They let me through, then I was flagged for extra attention before leaving the airport. I was not surprised.

The man wanted to know where I was staying. I said that I did not have a place, but that I would find one. I showed him my books, and everything changed, as is usual.

I was shaggy in appearance, my beard unkempt. I wore my magical vest as I call it, an amazing donation by the Mast General store, made by Mountain Hardware. Quality gear is well worth the price, being made to last a lifetime. It is near impervious to wind, with windstopper on it. The vest has zippered pockets, which are critical to avoid losing things. The air pocket created in storing gear creates amazing insulation at my core. I took off the vest, and packed it: now being in tropical Cuba.

The security officer apologized, but said that it was my appearance. I said that I understood, and that I was a veteran traveler, and that vests are critical to my travel so I can secure valuables while I slept. I said I would be just fine. I smiled a lot.

After studying my books, and my passport, and drivers license, he finally let me through. I got a pamphlet on where to stay, and realized that this would not be as easy as I expected. I thought I could find a much easier low cost place to stay.

I tried to get on a bus to a hotel. I had met a guy on the plane, who I had shown my books to. He came off the bus, and said that I could visit with him at the hotel. This seemed to ease the concerns of the bus driver, and eventually I was allowed on the bus.

I got on. I looked into getting off at the first hotel, but it was a five star. I explained my interest in an inexpensive place.

I got to the hotel, which was indeed splendid, a complete paradise, but lacking any natural quality. It was 115 cukes for the night (about $140).

I was scared: I thought I could stay a lot cheaper. I bought some sun lotion for about $20, a map, and tried just walking out of the hotel, past the police officer, with a smile. I had camping gear, I just needed to find a place to camp.

After a short distance, I walked into a Cuban military base, while I was looking for a place to camp. A police officer escorted me back to the hotel, and I got in a taxi, asking to go to a camping area. The first area was full, and then I went to Sitio La Guira, where they had a place. They said to look at the room, and if I liked it, I could get it for 20 cukes. It looked great, but I did not notice the water on the floor in the bathroom until later.

I paid for the room. I got into my room and was overjoyed. I had found a way to stay on Caya Coca within my means. Food and beer were reasonably priced, and this was a place of deep Cuban history, with many animals that I enjoyed being near. Oxen, horse, roosters, hens, turkey, two small very friendly dogs wandered free.

### Chapter 23: Sitio La Guira

Flowers graced the walkways, which were made of concrete blocks, but were quite aesthetic. Workers kept the vegetation from growing too high between the blocks using a hoe. I pondered whether a solar powered robot or vinegar would do this, but then the people would not have as much exercise, a key ingredient to optimal health.

A band played music, very nice, with a guitar, and sold their CDs for 13 cukes.

I took a walk toward the beach. I walked along their road, to the big green smiley face that graced the highway, with a green chair that served as a bus stop. I found a path/road leading north, toward the beach, and enjoyed getting off the highway.

What a wild day in Cuba. I walked from the Sitio La Guira where I stayed, out to the main road, then found a path through the forest going north, toward the beach. Piles of resources lay along the side of the road, and there was also lots of trash. But the forest was nice… much nicer

than walking along the road. I came to a large tower, and a fence. I tried

walking right, then came to a fence. I turned around, then walked up to a

security gate, and they let me in to the beach "La playa," they said,

Spanish for The Beach.

The hotel was very nice, with flowers, a flag of Cuba, and a nice

swimming pool. I walked to the beach.

A small bar was right on the beach, but I had already had two

beers. I wanted some AA batteries to get some photos, so I turned back,

asking people where the hotel store was. There was a guy selling small

ceramic statues out of a wheel barrow. I went by the exquisite pool, and

to the store, but it was closed, this being Sunday.

I went back to the beach, then went left (East).

The beach was quite nice, with lots of white sand, and young

woman wearing very skimpy bathing suits. I came upon a sailboat, and

dreamed of going sailing again. I took a full summer, every day all day

of lessons in Columbus of sailing when growing up. I knew how to sail,

though it might take some time to recall all the details.

I asked the person who was coming in whether I could go out sailing. The winds were brisk, and it would be an excellent time. He said that I needed to talk with the owner. I did, and filled out a form, but I did not recall the name of the place where I was staying, though I did know its location.

I realized later that I could have just pulled out my map to get the name. Instead, I continued to walk, for several kilometers, passing several young women. I was thrilled to be walking barefoot on the beach, with the soothing, archetypal sound of the crashing waves. This was indeed paradise. Cuba seems to have protected their beaches somewhat, keeping development back, and out of view. Much of this island was a natural area, though it seems nearly all of the giant trees had been cut down.

## Chapter 24: Rafting in Cuba

I walked into the forest, and noticed some brackish water.

I continued on, then decided to raft back, as the wind was going in that direction. I blew up my raft, and launched out. A couple from

Quebec walked by, and I told them about my brain implant, asking them to take it to their officials. "They probably already know about it," he said. "I think that Canada is doing it as well," I said.

I tested the wind, to make sure that it was not blowing out to sea. I was thrilled to be in the raft, though I was missing a piece of my paddle, so I had to paddle rather awkwardly. I wondered who took it, and when.

The water was very shallow, and I had to paddle several hundred meters out to get to water deep enough to paddle safely through. One shell could easily rip a hole through my raft, especially with the waves moving the raft up and down. The wind moved me back toward the exit from the beach, and I stayed in water where I could jump out and wade in should I need it. I greatly enjoyed the rafting experience, though I realized that I had not put my wallet in the dry bag, and it got soaked. I put it in the dry bag while larger waves splashed over the raft.

I paddled just to stay in the proper depth, going into shore when the water got deeper, then out when it was too shallow. I went past the beach, finding it amusing that I was rafting through where people were standing in waist-deep water – sometimes even shallower.

When larger waves spilled over the raft, I went on into shore, surfing the waves and moving surprisingly fast with little effort. On shore, I dumped out the water, and inflated the raft. The winds and wet raft chilled the raft, making it deflate some. I was worried it might have a hole, and decided to raft farther to make sure it was alright.

The beach was very irregular in nature: excellent for fish habitat. Where it was deeper (just about a meter) there was lots of vegetation. Sand bars that got less than a foot deep were a turquoise color, and I tried to avoid them, because of the chance of ripping a hole through the raft with sharp shells.

Luckily, the raft was OK, and it was very nice to raft farther, with the waves increasing in size. I turned into the waves, leaning back when they hit to help the raft surf over them rather than crash into them, which causes water to go over the raft.

I landed, and dried my gear, worried about thieves. A group of four people came by, with a young woman getting too close to my drying gear for my comfort. I think they thought I had them for sale, when I was just drying them in the hot sun. After drying my gear, I packed up and walked on out.

I found my way to the gate, with some effort, the hotel being like a maze. The security officer asked me where I was staying, in Spanish. I said Hablo Ingles. He said "finished" and I nodded my head, and walked out.

## Chapter 25: Lost in Cuba with Little Water

I tried to find the road where I came, and went on a path, leading north. I ended up getting very lost. The first path ended in dense brush, and I turned around and went back to the road. I tried another path, but it too continued to peter out. I thought that surely I must hit the road if I just continued on north. Instead, the paths kept getting smaller and smaller. I lifted a piece of wood up in the middle of the path to help mark where I took a left. I should have turned back: a big mistake.

I began to get very worried. The path ended in deep brush. Still hoping I was near the road, I continued on, and soon I was lost, in deep brush. I tried to go back, but it was getting so late that I could not even tell which way was west. A GPS would sure have been nice, but the government destroyed mine.

So, here I was, stranded in the wilderness. I got my headlamp out, and continued to try to find my way out. I was in deep trouble. Nearing dark, I finally laid down on a thick carpet of leaves surrounded by a volcanic formation. I switched between covering myself up with my jacket to keep mosquitoes off, and taking it off to avoid sweating too much.  Dehydration was my main concern in terms of surviving until I found my way out.

Without cell phone coverage, I was really in trouble.  The brain implant only made my survival worse, with their noise making it harder to determine whether noises I heard were people, such as the hotels, and where I should walk, or whether it was the thoughts of government thugs back in the US, using a cochlear implant making it harder to figure out what noise was coming from where.

They also continued their intentional infliction of emotional distress during a particularly troubling time.  They gave me a false sense of security, which caused me to just continue on when I could have easily turned back.

I laid down for some time, with mosquitoes buzzing around.  I considered my options, trying not to panic, which I know only made matters worse.  Mosquitoes were intense, and I worried about malaria.

If I waited until morning, with the forest being very young and thick, it might be cloudy, and I would not know where to go.  Going south would lead me into a thick wilderness, and I would likely go in circles without a compass.

The stars gave me the direction. I wish I had studied them better, but luckily, I did know a good bit. But it was hard to see them in the forest. I studied the map, and decided that if I went north, or east, I would undoubtedly hit a road or the beach, and be alright. South led into a large wild area. I figured Venus would be east, and I noticed it. I also heard an aircraft, and studied where the airport was.

My headlamp kept going out every several seconds, occasionally a minute. I had one extra AAA battery, and I replaced one of them, luckily the right one. The light stayed on. I turned the light out to rest, trying to keep from sweating too much. I only had a liter of water.

The forest was exceedingly difficult to get through, without a path. But I continued on north, actually backtracking for hour after hour.

Through the dense forest I walked, sometimes climbing up over slippery trees, using extreme dexterity to get through.

**Chapter 26: Hotel Music Saves the Day (or Night)**

After a couple of hours, I began to hear what sounded like music from a hotel. I was overjoyed, but concerned that it could be the nanonazis through their cochlear implant. Luckily, it was not. I asked for help, but they just laughed. I called for help from time to time.

I continued going toward the noise – northwest. Because north seemed the safest option, I went that way, until the noise was mostly west. I then turned west.

I got into very heavy forest, so thick it was near impossible to penetrate. I came upon a very large animal, with its eyes more than a feet across. It stamped through the forest. Luckily, it was also on a trail, but I was concerned about it charging me, and I stayed still, waiting for it to rush off some distance away.

I came upon the game trail that the oxen was using, and moved quickly through the forest. I tried to go toward the light and sound of the hotel, but the path went parallel to it rather than toward it. I had to go back into the thick bush. I did this, having a heck of a hard time, climbing through trees, and barely making it through on my hands and knees. I came to thick grasses. I had to hand my red Sierra Club backpack through the thick forest at times, then crawl slowly through, scraping against the brush all the while. The brush wanted to rip off my headlamp, which it often did.

The headlamp made it near impossible to distinguish the lights coming from the hotel. I was extremely hot and sweaty. My near empty water bottle was my main concern for making sure I got out safely, I often stopped, and turned off the headlamp, lying down on the forest floor.

I would then work to make out the stars to guide me, and notice the direction of the light and sound of the hotel.

I came upon another game trail, which led to a larger gravel path. Finally, it seemed like I was out of the dense forest, and I moved quickly toward the hotel. It was still a couple or more kilometers before I came upon the hotel gate, and walked by the empty security guard post.

I was soon greeted by security guards, though, who allowed me to purchase an internet card and use the hotel's internet. After checking my email, I next showed the security personnel my x-rays, and asked about seeing the physician.

I did see the physician, and she said that my x-rays were not normal. She gave me pain medication (iboprofin), but that was about it. She suggested going to a hospital. There was a woman from Quebec, and the Cuban police witnessed it as well.

I was surprised at each of these occasions that they would not contact the United Nations, and start an immediate, major international tribunal to investigate and prosecute the perpetrators. But, I was just sent away with pain pills, and the police drove me to the end of the hotel gate, asking if I wanted to take a taxi. I said that I would walk, having a bit of a challenge communicating because they did not speak English.

We rode out to the end of the driveway in complete darkness, as my eyes had yet to adjust, and I wondered how the police managed to see well enough to drive. I thanked them, and walked on toward Sito Liguero, about 5 kilometers away. I arrived, and was relieved. I ate a nice fish dinner, and spoke with the Canadian who had given me a ride the other day.

## Chapter 27: Playa Pillar

I organized my stuff, and got to bed early, planning to go out the Playa Pillar and go fishing. The next morning, I waited for the bus, and talked with a Cuban for quite a while. We each took turns trying to learn each other's language. We got a ride with a police officer and a military officer until they turned off the main road, then waited for another ride.

The Cuban at one point pointed to my book at the word heroin. I said no, no, and shook my head, saying that it was very bad. He seemed like an investigator, either with Interpol or the Cuban police. We got a ride with a Chilean, until the Cuban reached his destination, and then we went all the way to Playa Pillar.

I showed them where to park, then walked out, past the bar/restaurant, to the building housing a charter fishing company. I spoke at length with three people, one who understood English. He said that I should go in the fall, and that they had a tournament in October.

He offered to take me snorkling and over to the island where Hemingway frequented for free, even though it was usually 16 cukes (about $20).

I walked the beach, and had a very nice chicken lunch with coffee.

I caught a ride back with the same two people I rode out with, and they took me all the way to Sito Liguero, picking up two Cubans on the way. We packed in and it was warm, but soon the AC kicked on.

The Cuban in the front passenger seat was Cuban and spoke fluent English. The Chilean driver spoke broken English and the two added hitch hikers just spoke Spanish, so we had a fun time communicating.

I explained my brain implant, and showed them my x-rays. The Cuban said that they would help me out, but I would not see them again for the rest of the trip. But their ride sure did help me out.

## Chapter 28: Swedish Charm

When I returned, the band offered to play me a song, but I declined, saying that I needed to write. I ordered a shot of whiskey, and a beer. I ended up with two shots, and did not complain. I turned around and noticed two gorgeous young, blond Swedish women, around college aged, and stunning in their beauty.

I asked if I could join them. The eldest of the two, seated to my left on the open patio, said that I could. We began a long conversation, and I was pleased with the company of two very attractive and intelligent young women. We had a nice conversation about Cuba.

I was glad that they spoke English, and I did not have to work like the dickens to try to understand and communicate. My brain was stressed with using my dictionary and my studies of Spanish more than 20 years prior to try to communicate and not offend with so many Cuibans.

They had toured Cuba in great depth,the youngest, wearing a revealing white tank-top, pulled out a map, with the edges of it ripped off, likely from all of the extensive travel. They had circles on nearly all of the major cities, in all parts of Cuba. Like most tourists, they rented a car.

They taught me how to stay at homes for 20 pesos a night, showing me the upside down anchor sign, in blue. They said that the red was only for Cubans.  They said that people will ask for more money, but to not pay more than 20 pesos.

They said that the US had gained control of Guatanamo Bay prior to Castro's rise to power.

They were headed for the Cuban mountains next.  They asked where I was staying, and I pointed to my hut.  I considered inviting them into my place.  I had a giant bed, that could easily sleep three people comfortably, being at least 4 meters across, and I sure would have enjoyed their presence.  But I did not want to compromise my security, and passcodes while I slept.

The youngest of the two, who was sitting to my right, was drinking a bottle of water.  I pulled out the same type of water bottle from my red Sierra Club backpack, that I had purchased out at Playa Pillar, and drank some of it myself.  It was a warm early February day.  The young woman spilled a little water on my Arctic Melting book, and some on Against All Odds.  She wiped off the water on her white tank-top, while I wiped it off of Against All Odds, and shrugged, feeling my Arctic Melting book had just gained in value at such a tantalizing contact.

The same woman itched her legs and complained of the bugs. I said that lemon Eucalyptus is great for keeping them off, and good for the skin. I had forgotten to bring mine, but the bugs were of minimal nuisance to me, with my Arctic experience, at least thus far. They took off shortly thereafter, and for hours that evening I lay in my giant bed, pondering the exciting day.

## Chapter 29: Inside a Cuban Police Station

The following day, I hitched a ride in a large water truck to the police station about 8 kilometers toward the mainland. It took two rides, and it was a hot, sunny day. I walked in, and tried speaking in Spanish. Not a single one of the nearly dozen officers spoke English. I sat in a small cubical, where I showed them my IDs, and my x-rays, and they called an interpreter over.

I explained that I wanted to file a complaint against the United States, because the US was thought torturing me, and violating Cuban laws, as well as International laws. He translated, and they wrote down my complaint, writing my identification information from my passport.

They suggested going to the international health clinic a few kilometers away near a hotel. I got a ride to the road, then walked a few kilometers to the health clinic, getting a ride on a small diesel people mover on the way.

I walked into the clinic, and showed the man my x-rays, saying that I was tortured by the US government. I also explained how the US would not let me see a doctor, or to fix my hernia. He wrote a recommendation to see a neurologist in Moron, which was a few dozen kilometers away, across the causeway, on mainland Cuba.

## Chapter 30: To Mainland Cuba for a Hospital

I longed to be out fishing, and had had a reasonable offer to go out on a boat by the owner of Sito Liguero. But my number one concern was my health, as that dictated how many more fishing expeditions I could go on, and how many more books and films I could create, and how much of a difference I will make in the world. I needed an accurate diagnosis of my own health. What a sad state of American history when

I must go to another country to get an accurate diagnosis of my own health.

I packed up the previous night, then finished in the morning. I ate a hearty egg and toast breakfast, along with coffee.

I went to the road, and tried to hitch a ride. A taxi pulled up, and I reluctantly approached the driver, and asked how much it would cost to take me to a car rental place. He said it would cost 10 pesos. I said OK, and hopped in.

At the place, I handed him a ten peso bill, and went into the booth. It would cost 65 dollars a day, not including gas. He said he would rent it out by the day and take me back to my hotel, for 85 pesos.

I said that it was just a little over my budget, and that I would hitch hike. I had had great success at hitch-hiking, and money was very scarce where I live in southeast Ohio. I went out into the scorching sun, and tried hitching for a while. When I spotted a police station, I stopped trying to hitch for a while, until I got out of view. I was not aware of Cuban hitch-hiking laws.

I paused in the scarce shade in the mid afternoon sun to put on sun lotion. That lotion was well worth the price, because without it I would likely have taken a rental car, just to keep from getting sun-burned and the added skin cancer risk. Still, most sun lotion is somewhat toxic, and should be promptly washed off after the sun has gone down.

It was some time before I got a ride in a pickup-truck, which dropped me off near a gas station. A police car was on the road, with a police officer watching closely, and the driver look nervous.

I got out, thanking him, and went over to the gas station. I began walking around, saying I would pay for a ride to Moron, which was only a few dozen kilometers away, over the causeway in Mainland Cuba. After asking numerous people, the woman whom I first asked, in a very small car, said that it would be great. I managed to fit my large pack in the small car without problems, it being a soft framed pack, and got in the back seat with a Canadian woman, with her husband being in the front seat.

The woman said that her son and daughter lived abroad, and did not want to come back to North America. The causeway gave breathtaking views of the channel between Caya Coco and the mainland Cuba. It was far across the causeway, nearing the mainland when I noticed two wind generators, one of which sat idle. They were the full scale utility grade generators. She took a photo for me of the wind generators, using my digital camera.

As I had been told, once we got into mainland Cuba, I could see the poverty in which most Cubans lived. But I found later, that though the houses are built next to one another, they are actually quite nice and cozy.

They are made with much masonry, and they make it very easy to meet all of one's needs within walking distance. With much use of horses, bicycles and bicycle taxis, this is a very low greenhouse gas emitting country, and one very much at risk from climate change.

## Chapter 31: Inside a Cuban Hospital to Diagnosis US Government Terrorism

They dropped me off near the hospital. I asked where it was, and the driver pointed toward a building that looked like a run down school, with many open single paned windows. This was a crowded place, and a guy greeted me from an open window.

I walked past a bicycle parking area, which was large and appeared well used, and into the very crowded hospital. I went to the main counter, and handed over a referral to see a neurologist that I got at an International Clinic on Caya Coco.

They asked for me to sit down, saying it would be a few minutes. I sat down, and pulled out my Spanish-American dictionary, which I found was a great way to use time that otherwise would have been wasted, such as waiting for a bus or ride. It is so small, that I could carry it in my pocket, and pull it out easily.

I tried to pick up phrases from the many conversations going on in the crowded hospital waiting room. The doctor put a device next to a young girl's head just in front of me, and looked at a flat screen instrument. After about an hour, the place mainly cleared out, and I again approached the main counter. They said it would not be that much longer.

About a half hour later, someone came down, and had me follow him up to the neurology area. I was taken to a waiting area. At the corner of the building. Spotting a 110 outlet, I got out my laptop and began writing. It is always so hard to keep up with writing when traveling, because so much happens. I was always so busy with the chores of travel, like packing, and finding my route, that writing is usually the last thing on my mind unless I prioritize it. Also, finding an outlet, and an adapter was hard to do.

After about 10 minutes, three people came in, Dr Angel, another male doctor, and an attractive 21-year old female med student. I showed Dr. Angel my x-rays, and he said that he wanted to see my MRI images. I pulled out my two portable hard drives, but said that I did not have the cable. He said to come back the next day, with a cable, because he needed the MRI images. I asked if he knew of a place to stay.

He made several calls, using both his cell phone and the hospital phone, and wrote down the address. The female medical student, said that I should not pay more than half a peso for the pedal taxis.

We had quite some time to talk, while the two male doctors were out, and I enjoyed the talking with the med student. She said that I could buy a home in Cuba. I wondered how much it would cost, and how that would work in a communist country.

## Chapter 32: Affordable Luxury Lodging

I found my way out of the large, winding hospital with several floors, and taxis of all types, pedal, horse and auto were waiting. I gave the note to one of them, and asked if I could get there for half a peso. Another pedal taxier followed, and asked if I wanted a Chica, a woman. I smiled, but said no, that I did not. I did not want to catch a disease or waste money, or engage in what I consider immoral behavior.

He continued to pester me the whole way, nonetheless, then he even tried to get me to pay him just for his solicitation. I gave a peso to the bike taxi pedaler. It had been a very short ride, well under a mile. I asked for change, but he took off.

I went to the gate which had an upside down blue anchor sign, signalling a place where tourists could stay. The woman said she was full. She pointed across the street. I knocked on the door of an exquisite house, and was shown a spectacular place. I tried to talk him down, but he would not go under 20 pesos. The place had a nice balcony, and patio.

I gave him 40 pesos for the two nights, and signed my name and contact information in a book that he had. It was a very nice place.

He asked if I wanted dinner. He said he could make pork, chicken or fish. He said that they have breakfast as well. I asked if it was included in the price of the room. He said no. I said I would find food out in the city. I wanted to wander, I was very good at finding bargains. I did not want to sit down. I like to eat while walking.

I unpacked, changed out of my t-shirt, and washed it, and a pair of socks. I went outside, and enjoyed what the owner of the house I was staying at called The Mall. Shops were everywhere, and I soon found where to get food and beer.

But my first priority was to get to the telecommunications building, and get a cable for my portable hard drive. I went there, and got a rude reaction, though I did not know Spanish well. Eventually, I was taken back, and I saw the cable I needed in a windowed cabinet. I pointed to it. It plugged in alright, but did not open up on my computer. I still tried to buy it, but he would not allow me.

He pointed to a video place, where I went, and I was offered the cable for 15 pesos. I talked him down to 10, knowing he could get one for 4 pesos a block away. Annoyed at the waste of 6 pesos – about 8 dollars, I went back to my room, and got my computer working. I was able to get the portable hard drive working, and moved the MRI images to a flash drive. I then went back out on the street to eat and drink beer, and explore the city of Moron.

There were two currencies used in Cuba, a local currency, which tourists are not supposed to use, though I was told by a Canadian that it was fine to use for tips and food, and the Peso, called by local cukes, which is worth 20 times more, and is used in more upscale shops and restaurants geared to tourists.

I walked around, trying to change a twenty dollar bill, but the banks closed a 3:30. A guy said that he would change it. I was a bit reluctant, but went along with it. He took me into his house, and said that he did not have the change, but that he could go out and get it. We were on a main street in town, near the train station. I got the local currency, about 380 pesos for 20 US dollars.

After I began buying three peso ice cream cones, 20 peso beers and 10 peso pizzas, I felt like a rich person. My money went exponentially farther using their local currency. I later found 10 peso cheap beers, at the same place that sold 10 peso pizzas. Because I was walking usually more than a dozen miles per day, I needed a lot of calories. Climbers of Mount Everest burn up to 10,000 calories per day.

I ate a lot, because there was so much, so cheap, so good. Because good food had been quite pricey on Caya Coco, I conserved on meals, though ate plenty. I had also done an enormous amount of walking, and some paddling, and had an explorers appetite. I had an ice cream cone, managing to drip chocolate ice cream on my pants, which I later washed off in the sink in my very nice room.

Crowds of people waited for buses and trains at the train station. I saw the worker bus, as the locals call it, which would be called a third class bus in Mexico, with people crowded on so thick that it reminded me of getting on a downtown Tokyo Subway during rush hour, which I did several times during my 5 week speaking tour of Japan in 1993.

A line of horses and carriages, taxis, bike taxis, bike rental, car rentals and buses were all around the train station. This is the kind of integrated transportation that we need everywhere, where one can go from local transportation to cross-country transportation at the same place.

Many shops sold right out onto the street, and people stood around, eating and drinking, then returned the plates and cups when finished. Very few disposable products were used. The cobblestone streets were very narrow, European like.

They were narrow, crowded walkways with much bicycle and horse traffic, and the occasional bus and taxi that people cleared the way for. Watching buses push their way past nonchalant bicyclists, with just inches to spare, I was relieved to be walking, even if it was slower. It also made it easier to take photos.

I got the cable TV working, and found an English speaking movie, with subtitles in Spanish. I found this a very good way to re-learn the language (the movie Merlin, quite comical). I also watched the news (in Spanish), finding that I could comprehend a significant amount, though I needed a lot more studying before I was fluent. I decided to cut off my beard, initially using some borrowed scissors, then several raisers that I had bought in Toronto. It was a long, uncomfortable process. I wanted to make a good impression on my neurologist.

The next morning, I went out, and got one local peso (about 6 cent) coffee, and three peso cheese sandwiches very close to my hotel. I also had three sweetened expresso shots from another vender.  I paused at a barber shop, but it would not open for half an hour, and there were already three men waiting.

I got some good photos of a horse pulling a cart.  I paused by some men lifting a refrigerator onto a horse drawn cart.  While some might call this old-school, or third world, I find it a very pleasant atmosphere, provided the horse is well treated, and not overworked or over-heated.  I had seen a horse that looked so hot, being completely covered in sweat, in the sweltering early February heat on Caya Coco.

## Chapter 33: Documenting the Brain Implant with a Flash Drive

I got to the hospital at about 8:15 in the morning, having told them I would be there by 9.  As it turns out, I should have waited until 9. I just sat and waited, eventually being taken up to the neurological area, where I waited.  I studied Spanish for about an hour, before being asked to pay 25 pesos to be admitted.  I paid, waiting quite a while for the change.

I gave my MRI to Dr. Angel, which he reviewed. Some time later, I went back to the corner room, and talked with the same doctors and med student as the day before. Dr. Angel said that I had a fracture in my frontal lobe. He said I had a skull fracture linked to the frontal sinus communication and a brain implant. The diagnosis was that I had a brain implant in my sinus cavity engaged in communications via satellite.

He wrote this down, both in Spanish and in English. I asked about sticking a camera in my nose. He said that would be possible, an endoscopic procedure. He also suggested a CT scan, which I said that I was reluctant to do because of the amount of radiation being very significant.

He said that because I was leaving the next day, Saturday, that I should wait to come back to do the endoscopic procedure. I said that I should be able to come back shortly. The woman suggested that I talk with FAR, the Cuban military. She suggested that I not bother with the police, as they probably would not have the equipment to test for wireless frequencies.

I got the diagnosis of a skull fracture, and frontal sinus communication and brain implant in both Spanish and English, which I would later post on the internet.

Upon leaving the hospital, I found a gargantuan tree, about 10 feet in diameter, with trash littered upon it, and two calzone places on either side of it. I thought it was such a disgrace to treat the only large tree that I had seen in Cuba in such a way. I took several photos with my digital camera, and went up and touched the tree in a loving way. I couldn't resist a little hug.

I bought a calzone for 5 pesos, about 30 cents. I then stopped by a very grungy looking restaurant, with unclean tables, and one guy sitting at a table, complaining. I bought a beer for 18 pesos, and a hot sandwich with a fried meat concoction for 2.50 pesos, about 13 cents. It tasted fine.

On down the road I went, and across the railroad tracks. I wanted to explore some. I did not realize how big Morone was, and it is a very densely built city, though I found it very attractive and enjoyable, with few exceptions. Nearly everyone was exceptionally nice and patient, though few spoke English and even fewer spoke it fluently.

At one point I asked for the train station, and getting blank stares, I pulled out my Spanish-English dictionary and asked for the Estacion de tren. They pointed the way, saying it was about a kilometer. I walked by someone making mortar in a donut shaped mound of dry mortar mix, with the water poured in the middle. Having done a lot of masonry myself, I was fascinated with how they did things.

I walked on and peered in a hardware store. At the square, I went in to the telecommunications building and went into the building, to the room where they would not sell me the 4 peso cable. I asked if they would test for wireless frequencies. There were two women and a man, well dressed, with older-looking tower computers. They had big smiles, and I pointed out my x-rays.

Sighing over another failed attempt to expose the thought torture and terrorism that I was a victim of, I walked out, in the mid-day, sunny day, along the path through a narrow square, with an old woman sitting on a bench. I stopped by a corner store I had already been to several times, and bought a beer and a large hotdog. I try to avoid processed meats, but there was not much choice.

Later in the evening, I toured the train station, which I found quite pleasant. Across the small square, I noticed several people hanging out at a storefront. Like many in Moron, it was just an opening into a building, where they served drinks and meals, and people stood around eating. Here, I found healthy food, with beans and rice and salad, along with juice, all for about a dollar, because I was using the local currency.


## Chapter 34: The Long Journey Home

The next morning, I got coffee and a sandwich. I asked about getting to Caya Coco, if there was a bus going there, in Spanish. The man knew a guy going there who happened to be walking up to the shop. I showed him my three books. He spoke English, and said that if I hurried, I could catch the tour bus. I did hurry. Initially they would not let me on, then the guy talked to the driver for a little while. The driver let me on.

It was like a scene out of a comedy as I tried hard to keep my giant pack from whacking people as I walked through. Several Cuban police were at the back of the bus. It was a challenge to find an empty seat, and a place for my giant pack. I was dropped off at the road leading toward the airport, and hitch-hiked in. I had planned to raft around near the airport. I got some coffee and a hamburger at the tiny shack near the Caya Coco Airport.

I chatted with the tourist guy, and showed him my book. I paid a taxi driver, Rauel, 25 pesos to take me to the beach and back in time for my flight. He insisted that I sit in the front seat. I enjoyed a beer at the beach, writing this narrative.

If this was not paradise, I did not know what was. Sea food was available, and beer, in a spectacular, natural covered patio, with a boardwalk walkway leading to the beach. Sailboats came by. The waters were a turquoise paradise color, though I wondered about the remnants of the BP oil spill.

I took several photos. The water was out, so I used some Cuban tap water that I had put in a water bottle, and biodegradable soap over the edge of the walkway. I knew it would just serve as fertilizer.

The taxi was waiting for me, and I threw my giant pack in, and we sped back to the airport. I had to pay another 25 pesos before they would let me into the waiting area: the airport tax. Luckily, a money exchange place was available.

I went through all of my gear, searching for pesos to spend before I left, always a pleasurable time when leaving a county, as it is cheaper to buy things rather than pay exorbitant exchange rates that most places charge. I got a cuba magnet, and pondered getting some books on the CIA and Cuba. But I decided I could save money and get them used later. I bought some fried chicken and potatoes at the restaurant in the waiting area, and talked with the guy who I had met on the plane coming to Cuba. I had shown him my books.

He had helped me get onto the tourist bus to get me out of the airport. He said that he went back to clear off some snow from his automobiles, which I thought odd. I met his wife, and his children. I went through some of my major adventures. He seemed envious, and mentioned several times about how children tied him down.

My dog, Dude, was a new tie for me. I had to pay $12.50 a day to house him in a kennel, and feel bad the whole while that he was not treated as the King that I treat him as.

I flew back to Toronto. There I got on a subway towards a hostel I booked at the airport, and typed very fast, typing about this journey. My Mom taught me to type at an early age, and I can type 120 words per minute. I think practicing piano for 12 years helped.

My journey was not only so exciting, particularly how I did it on such a low budget, it was also very important too. I got a neurologist to give a diagnosis that I not only have a brain implant, but skull fractures too. This proved I was right, and doctors in the US are lying. You can see that diagnosis and x-rays yourself at www.brainimplant.info

Now I face an uncertain future. I need the help of the global community and every US citizen to demand that the US admit publically to what they have done to me, and compensate. You do not need to be a doctor, or a world leader to take action.

I have begun a new nonprofit organization: Citizens Concerned about Brain Implants. You do not need to support me personally to be a member: you just need to help me get the truth about what our government has done to me. This can help humanity for all time, as this technology continues to expand, and become easier to administer against people's will.